Jan goes to see Gran.

Gran hugs her and says "Hi!"

First, Jan wants to see Raffle.

Then she and Gran put corncobs in the hogs' pen.

"What cute little chicks," says Jan.

They pick apples.

Does the basket look full?

Gran starts to juggle.

"Your turn!" says Gran.